One-Matter
NATURE SCIENCE

One-Matter
NATURE SCIENCE

Tsau's Scientific Revolution

Josef Tsau

Copyright © 2022, Josef Tsau

All rights reserved. Printed in the U.S.A.

No part of this publication may be reproduced or transmitted in any form or by any means, electronic or mechanical, including photocopy, recording or any information storage and retrieval system now known or to be invented, without permission in writing from the publisher, except by a reviewer who wishes to quote brief passages in connection with a review written for inclusion in a magazine, newspaper or broadcast.

Quantity Purchases:
Companies, professional groups, clubs, and other organizations may qualify for special terms when ordering quantities of this title. For information, email info@ebooks2go.net, or call (847) 598-1150 ext. 4141.
www.ebooks2go.net

Published in the United States by ebooks2go, Inc.
1827 Walden Office Square, Suite 260,
Schaumburg, IL 60173

ISBN: 978-1-5457-5560-0

Library of Congress Cataloging in Publication

Contents

Preface -- vii

**Chapter 1: Summary of Tsau's Breakthrough
　　　　　　 Scientific Discoveries** -------------------- 1

　A. (Nature) Science Having a Specific
　　 Definition-- 1

　B. The One-Matter Nature Science------------------ 1

　C. Tsau's Scientific Revolution ----------------------- 2

**Chapter 2: Rediscovery of Science and the Long
　　　　　　 Chaotic Era of Science** --------------------5

　A. The Loss of the Definition of Science to
　　 Scientific Debates --------------------------------- 5

　B. Rediscovery of Science---------------------------- 6

　C. Tsau's Struggles to Save Science ---------------- 10

**Chapter 3: Tsau's Breakthrough Scientific
　　　　　　 Discoveries** ------------------------------ 13

　A. The Charge Energy of Electron and Proton----- 14

　B. Matter Having Atomic and Molecular Structures
　　 and Universal Phenomena ---------------------- 18

　C. Light Particles ---------------------------------- 20

　D. Cosmic Microwave Background (CMB) -------- 22

　E. Elementary Matter Particles—the One-Matter
　　 Nature Science ---------------------------------- 23

F. New Cosmology and Astronomy ---------------- 24
      1. The New Definition of the Universe --------- 26
      2. The New Definition of the Outside
         Universe and *EDM* -------------------------- 27
      3. The First-Generation Stars ------------------- 28
      4. The Second-Generation Stars ---------------- 30
      5. Galaxy Wind of Neutrino Particles ----------- 32
      6. Science of Gravity and Antigravity ----------- 32
      7. Stars Having Planets ------------------------- 34
      8. Possible New Kind of Matter Having
         Strongest Gravity ---------------------------- 34

**Appendix: The Story of Tsau's Scientific
   Revolution ------------------------------- 37**

   A. Legal Challenge of Scientific Correctness ------- 39
   B. Efforts To Convince Scientific Authorities ------ 49
   C. Time for the General Public to Be the
      Real Owner of Science --------------------------- 61

Preface

To start a scientific revolution, Tsau used this book to scientifically challenge the scientific community, including governmental scientific authorities. In the past two decades, Tsau has already repeatedly written to scientific authorities to show them that modern physics is not science needing to be removed from science but has been ignored. Hopefully some dedicated scientists of governmental scientific organizations will read and evaluate this book resulting in getting their approval and support scientifically. First of all, Tsau hopes that US government can timely explore his breakthrough scientific discoveries to enhance its worldwide leading role in science and technology. Since both saving and advancing science are the responsibility of everyone, Tsau hopes that the general public will read his book to understand what is and is not science. Thus, becoming the real owner of science to be able to make sure that our scientific community and our government are performing their scientific duties correctly. Tsau further challenges corporations and public organizations to participate scientific revolution to take on their responsibilities of saving and advancing science and encourages scientific and technological companies to explore and to utilize his breakthrough scientific discoveries.

This is the last book to conclude Tsau's investigational R&D on whether modern physics, today's mainstream of thought of science, is science or not and if not, what are the correct scientific answers. First of all, Tsau intends to show that science is logically understandable even by the general public and it has nothing to do with difficult mathematics. He teaches that the definition of *science* must be obeyed to be science, that the science discovered by Copernicus and Galileo also known as the mechanical physics–based science is the only correct science, that scientific community has long been disobeying the definition of *science* to develop and to teach modern physics, which is religion-like pseudoscience needing to be removed from science, and introduces Tsau's breakthrough scientific discoveries to advance science. This book covers all scientific topics without offering references, since they can be conveniently found on Internet such as in Wikipedia.

Unexpectedly devoting his early retirement to scientific R&D against the teaching of scientific community has brought him hardships. Yet Tsau has found it challenging and enriching to his retirement. He calls the science he has developed his family science, since it has often been the topics of his family discussions particularly with his son John, an engineer, and his diseased wife, Gertrude, a physicist. He expects that his family members will also continue to help to take on the scientific missions until science is saved and his breakthrough scientific discoveries are officially evaluated and utilized to advance science.

CHAPTER 1

Summary of Tsau's Breakthrough Scientific Discoveries

A. (Nature) Science Having a Specific Definition

Science has long been without a definition or undefinable. Tsau has rediscovered the definition of science showing that it has a specific scientific method, the experimental method, to make scientific discoveries and interpretations. He shows that the definition of *science* must be obeyed to be science. Accordingly, the mechanical physics–based science is the only and the entire science while modern physics, which disobeys the definition of science, is not science.

B. The One-Matter Nature Science

The nature can be interpreted by having only one kind of elementary matter having amazingly dense mass and its particles take up the space of the nature. In the universe they are produced by the nuclear reactions of all stars, thus, are energetic and their mutual collisions continue to break each other up to finer-size particles. Unstable to collisions to be breaking up, these particles cannot form matters, which are stable to collisions.

The nuclear reactions in stars produce, besides an atmosphere of energetic neutrino particles, which are undetectably small, both protons and electrons carrying charge energy. Tsau suggests that for a single particle to be stable from breaking up by collisions, it needs to be surrounded by a dense cloud of fine (neutrino) particles, which collides interacting with the atmospheric energetic neutrino particles to produce its charge energy. Having such a structure of composite particles, both proton and electron carrying charge energy to collide interacting with each other to form unlimited number of matters such as the matter having atomic and molecular structures. To explain the existence of lights, Tsau further suggests that some larger single neutrino particles have also been stabilized by surrounding with a dense cloud of finer neutrino particles to carry mini-charge energy to form lights but only inside matters such as matter having atomic and molecular structures. It is reasonable to conclude that all elementary particles are the same kind of matter and therefore it's an one-matter nature.

C. Tsau's Scientific Revolution

The science discovered by both Copernicus and Galileo has led to develop the mechanical physics–based science, which uses experimental findings only to discover and to interpret everything, thus, obeying the definition of science that science has a specific scientific method, the experimental method, to make discoveries and to interpret findings. With one specific scientific method, everything has only one specific experimental interpretation meaning that the definition of science must be obeyed to be science. Since experimental method can detect and study matter particles only, science has limited its scope to study matter only. Yet our scientific community has long been developing theories based on both postulations disobeying the definition of science and misusing mathematical-derivation method to replace experimental method resulting in developing and teaching religion-like pseudoscience known to be modern physics, which unfortunately has become today's "mainstream of thought science" covering such important scientific topics as particle physics, matter science, universal phenomena, cosmology, and astronomy. The event has been known to be Einstein's scientific revolution, which calls the original science the classical mechanical physics and teaches that it is incapable of explaining relativistic and quantum phenomena, thus unable or unacceptable to be used to interpret the scientific topics already taught by modern physics.

Tsau's scientific revolution shows both that modern physics is not science since it disobeys the definition of science and that mechanical physics–based science now can interpret everything detectable experimentally and different from modern physics, thus, is the only correct science.

CHAPTER 2
Rediscovery of Science and the Long Chaotic Era of Science

A. The Loss of the Definition of Science to Scientific Debates

Upon the discovery of science, the teaching of science and the conducting of scientific R&D have quickly become the major academic and industrial activities of all nations worldwide. Under great academic freedom, there have been long and hot scientific debates on whether lights are matter particles or massless waves and whether gravitational force is produced by tiny particles or not despite that the definition of science already had the answers. Unfortunately, in both major scientific debates the winners have been those disobeying the definition of *science* resulting in the development of today's modern physics.

God's cruel joke to match the theoretical speed of electromagnetic wave to that of light found experimentally has played an important role to mislead scientific community to develop mathematics-based modern physics believing that electromagnetic wave had been proven experimentally to be light. Yet electromagnetic wave is both massless and having an absolute speed or no relative motion meaning that it is not matter particle, thus, is not detectable experimentally. By definition of *science*, lights have to be matter particles to be experimentally detectable and therefore electromagnetic wave should not have been acceptable to be lights in the first place.

Upon accepting the conclusions of the above-mentioned two major scientific debates that light is massless wave and that there is no need of tiny matter particles in space to produce gravitational force, our scientific community has been misled to develop and to teach the mathematical theories (Maxwell's light theory, Einstein's theories of relativity, quantum theory, standard-model particle physics and big bang theory) of modern physics as science. Since these theories are based on both postulations disobeying the definition of science and using mathematical-derivation method, not experimental method, to make scientific discoveries, they are not science. For example, without having matter particles to produce the universal forces of the modern physics, they are not detectable or provable to exist experimentally or not science; therefore modern physics teaches a fantasy universe having its universal phenomena not really provable to exist experimentally. Our scientific community has been in a long chaotic era by its leading Einstein revolution to develop and to teach modern physics as today's mainstream of thought science to replace the mechanical-physics-bases science and to block its advancement.

B. Rediscovery of Science

Upon realizing that modern physics cannot be correct science, Tsau has dedicated his early retirement to search for real scientific answers to modern physics and to try to convince scientific authorities to do the same. Yet his repeated efforts to publish his findings in scientific journals have failed and his writings to scientific organizations such NSF, American Institute of Physics, boards of education of different states, United State presidents, etc., have not been responded. He has even brought both NSF and American Institute of Physics to federal court in Chicago more than once without getting a chance either to debate science in court or to get them to respond to his serious scientific accusations.

But Tsau has continued to be encouraged by getting steady progress from his own scientific R&D and eventually he has made the breakthrough scientific discoveries to interpret everything detectable experimentally including finding completely different answers to those of modern physics to, again, disprove that modern physics is not science agreeing with the conclusion made based on its disobeying the definition of *science*. If modern physics were correct science, Tsau's breakthrough scientific discoveries can never happen.

In summary, Tsau has discovered that mechanical physics–based science is the only correct science teaching that the space of nature is filled with the particles of one kind of elementary matter, which has such amazingly dense mass that the collisions of its particles continue to break themselves up to finer particles. Yet a few of them can be stabilized from breaking up from collisions by having the structure of composite particles of a single large particle surrounded by a dense cloud of fine (neutrino) particles, such as proton and electron, which have also gained charge energy from their collision interactions with the atmospheric energetic neutrino particles, to further collide interacting with each other to from unlimited numbers of stable matters such as that having atomic and molecular structures and many universal phenomena. All matters such as their heavenly bodies need an atmosphere of energetic elementary (neutrino) particles in space to constantly collide interacting with them to continuously supply the charge energy to their electrons and protons to form them, to keep them stable, and to produce universal phenomena.

A summary of experimental proofs of why modern physics is not science are given below.

- Einstein's mathematically interpreted universe: It is a constantly expanding universe having heavenly bodies in space-time, which can be curved by massive matter to produce gravitational force. The curved space-time or Einstein's gravitational force do not have matter particle to produce, thus, should be undetectable experimentally but the real gravitational force is. Besides, there is no energetic particle in space to collide interacting with its heavenly bodies to produce them and universal phenomena. Therefore Einstein's mathematically interpreted universe is not science.

- Mathematically interpreted universal forces: Newton's gravity theory postulates that gravitational force is the attractional force of heavenly bodies acting from distance. Its success has led to the postulation that all universal forces are such attraction force and the development of theories to interpret it such as force field theories. The existence of pushing force on contact of matter particles has long been experimentally proven to exist. Yet the attraction force acting from distance without having matter particle to produce it has not yet and cannot be proven to exist experimentally. Therefore it is not science.

- Einstein's theories of relativity: Since their fundamental postulations, such as that light were electromagnetic wave, disobey the definition of science, they are not science. For example, its relativistic phenomena, such as spacetime have no matter particle to produce them, are not detectable or provable to exist experimentally.

- Quantum theory: Based on the assumption that light is massless wave, the spectra of light produced by matter are expected to have continuous energy. Yet findings show that these light spectra are not continuous has led to the development of quantum theory assuming that light is quantized energy, thus, having turned light from massless wave to particle-like becoming scientifically correct. Therefore the classical quantum theory has made valuable scientific contributions, such as its statistical interpretation of light spectra to elucidate atomic and molecular structures. Yet upon adding the postulations of Einstein's theories of relativity, quantum theory became a famously weird theory and is no longer science.

- Big bang theory: It is based on the interpretation of the findings of universal redshift of starlight, which increases with distance, using Hubble's law based on Doppler effect teaching that the universe is constantly expanding starting from the explosion or "big bang" of an extremely hot and dense "singularity" without being able to explain how and why a singularity can form and exist in nature. Besides, there is another possible interpretation of the phenomenon: the presence of an atmospheric tiny matter particles to collide interacting with starlight particles in space. This scientific possibility is supported by the presence of an atmosphere of neutrino particles in space, the existence of heavenly bodies and universal phenomena but has not been considered by the scientific community.

- Standard model particle physics (SMPP): It is mainly based on the experimental findings of the collisions of charged particles in particle accelerators but it has not taken into consideration that the charge energy of all charged particles is unstable to fast motions and acceleration. As a result, it cannot find the scientific reason why particles cannot be accelerated to light speed. So many elementary charged particles have been found that they are called particle zoo. Yet the most important conclusion of the overall findings is that besides electron and proton, all other charged elementary particles are unstable in nature. Although the theory is based on experimental findings, it cannot be science since its interpretations are based on many postulations disobeying the definition of *science*. Its fatal mistake is its ignoring to find out what the charge energy of proton and electron is and where its energy source comes from.

C. Tsau's Struggles to Save Science

Again, if modern physics were correct science, Tsau's breakthrough scientific discovery enabling mechanical physics–based science to interpret everything detectable experimentally can never happen. It has taken nearly three decades for Tsau to go through steps of scientific discoveries and correcting existing scientific concepts. Despite that all his efforts to present his scientific discoveries to scientific authorities have failed, he is happy to be able to achieve his important scientific goals of proving that modern physics is not science and to make breakthrough scientific discoveries to replace modern science and to advance science as a whole.

Recently, he has launched a new round of efforts to introduce his breakthrough scientific discoveries to scientific authorities. Since our new president Joe Biden emphasizes science by raising its importance to Cabinet level, Tsau has written to him hoping to convince him and his science team to save our science. Upon failure of getting response, he again has written to governmental authorities, namely NSF, NASA, and Department of Energy. Again, no response has obtained.

Tsau shows that the definition of science must be obeyed to be science. Without realizing it, besides mechanical physics–based science, our scientific community has also been developing and teaching modern physics based on postulations disobeying the definition of *science* leading to a long chaotic era of teaching two contradictory "sciences." He therefore asks governmental scientific authorities to remove modern physics from science. Also, he wants to introduce his breakthrough scientific discoveries to advance science. Upon failed to get response from scientific authorities, he has written to Eddie Bernice Johnson, chairwoman of the Committee on Science, Space, and Technology of Congress, hoping that Congress can help legally protecting the definition of science to end the long chaotic era of science. Again, it resulted in no response.

Today's general public has long been brainwashed to believe that science only belong to a few rocket scientists understanding difficult mathematics. Fortunately, this is not true. Mathematics has always been the important tool of science such as using statistic calculations to quantify the efficacy of drugs and vaccines but difficult mathematics has nothing to do with science, which is beautifully logical and easy to understand. Humankind is born to be logical, curious, eager to learn new scientific knowledge. Hoping that this book will convince the general public that they can understand science, what is, and what is not science. He hopes that upon reading this book, the general public will become the true owners of science to be able to make sure that both our scientific community and government are properly performing their scientific duties. More detailed story of Tsau's to save science is given in Appendix I.

CHAPTER 3
Tsau's Breakthrough Scientific Discoveries

Upon finding that modern physics is contradictory with mechanical physics, thus, cannot be correct science, in the past three decades Tsau has been searching for the correct scientific answers to replace those of modern physics. Based on that correct science must obey the definition of science or that both matters and phenomena must be experimentally detectable, gravitational force must be produced by matter particles and lights must also be matter particles. To start with, he needs to understand what the charge energy of electron and proton constantly carrying is and where its source comes from. He also doubts that particles carrying spin energy without having an energy source. Upon realizing that the only scientific possibility for proton and electron to gain their charge or any kind of energy is from their collision interactions with the atmospheric energetic neutrino particles, he has not only discovered what charge energy is but also a lot more, such as the science of elementary particles, the important scientific roles the atmospheric energetic neutrino particles play, the science of matters, the universal phenomena, lights, etc.

Also, Tsau has discovered both new cosmology and new astronomy obeying the definition of science based on his bold postulations that the absolute zero-degree temperature is the transition point for matters having atomic and molecular structures to degrade by collapsing their orbital electrons into atomic nuclei to form a new matter called extremely dense matter (EDM) by Tsau stable at absolute zero and below temperatures. Consequently, Tsau's breakthrough scientific discoveries of the mechanical physics–based science has led to teaching a different science from what we have learnt and students are learning in schools and universities mainly due to that modern physics has long been the mainstream of thought of today's science but it is not science.

A. The Charge Energy of Electron and Proton

According to standard-model particle physics, electron is an elementary particle, and proton is made of quarks. But what is the charge energy they constantly carry and where its energy source comes from have not been addressed. The only scientific possibility for both electron and proton in space to constantly carry charge energy is that they are able to collide interacting with the atmospheric energetic neutrino particles constantly produced by the nuclear reactions of all stars to gain it. It also means that the matter having atomic and molecular structures made of protons and electrons also need to be able to collide interacting with the atmospheric energetic neutrino particles to keep their protons and electrons constantly charged to form them, to keep them stable, and to produce universal phenomena. Yet the scientific community continues to teach that since both proton and electron are tiny particles that matter having atomic and molecular structures are essentially vacuum space, the atmospheric neutrino particles pass through it such as Earth freely without having significant collision interactions to produce anything.

There is no choice but to research on the only scientific possibility that the atmospheric energetic neutrino particles do collide interacting with proton and electron to produce their charge energy. If electron is a single or tiny point particle, it has no way to utilize the kinetic energy obtained from its random collision interactions with the atmospheric energetic neutrino particles besides been breaking up to finer particles and producing heat. It means that electron cannot be a single particle but having some kind of structure of composite of particles. The R&D using particle accelerators has found a zoo of particles carrying charge energy but, beside electron and proton, they are unstable, likely, to be broken up by their collisions with the atmospheric neutrino particles. Therefore it is very unique that both electron and proton are the only large elementary particles carrying charge energy somehow can be stable from breaking up by collisions.

A possible natural way to protect a single particle from been breaking up from their collisions is by surrounding it with a dense cloud of fine particles and both electron and proton may be such a structure of composite particles. It is possible that all other charged particles also have this composite-particles structure but the clouds surrounding their large particles are not dense enough to keep them from breaking up by collisions.

The important question is that if both electron and proton have the composite-particles structure of a large single particle surrounded by a dense cloud of neutrino particles, besides been stabilized from breaking up, can their collision interactions with the atmospheric energetic neutrino particles also produce their charge energy? First of all, having such a composite-particles structure is a lot more effective in colliding interacting with atmospheric energetic neutrino particles than a single particle to gain their kinetic energy. The collision interactions result in constant gaining kinetic energy from the atmospheric energetic neutrino particles should also result in making it a low-energy center in space. It means that the atmosphere of energetic neutrino particles has an atmospheric pressure or inward-pushing force on it from all directions tending to push all charged particles towards each other. Meanwhile, the constant bombardments of the atmospheric energetic neutrino particles also cause the dense cloud of neutrino particles turbulent to produce an outward pushing force. The two forces are opposite in direction and, likely, different in strength and range from each other; therefore the charge energy of having such a composite-particles structure as an electron or a proton, is expressed by two opposite forces gotten from its collision interactions with the atmospheric energetic neutrino particles. To be scientific correct, this theoretical charge energy should be able to interpret all the interactions among electrons and protons such as to form matter having atomic and molecular structures.

Let's see if the above-proposed two-forces charge energy can interpret the collision interactions among protons and electrons or not. When a proton approaches another proton or an electron approaches another electron, they repulse each other. This can be interpreted by that having the same size, their outward pushing forces are maximized and are stronger than their inward pushing force. When a much smaller-size electron approaches a proton, however, initially their inward pushing force is stronger than their outward pushing force resulting in approaching each other until that they are so close to share part of the space of their dense cloud of neutrino particles to reach the balance of the two forces; therefore the proposed two forces charge energy can interpret the fundamental collision interactions among protons and electrons to form matters having atomic and molecular structures. It will further explain in the following scientific topics how the properties of the dense clouds surrounding both electron and proton inside matter having atomic and molecular structures be changed to have liquid-like properties by forming light particles inside them for their charge energy to perform such complicated functions as physical, chemical, nuclear bonding and reactions. It further means that protons, electrons, and the matters made of them have a high content of neutrino particles and this fact has long been proven correct by huge amounts of findings such as that all nuclear reactions and radioactive matters emit neutrino particles.

Finding of what is the charge energy and its source for both electron and proton has led to many important scientific breakthroughs. For example, it revolutionizes particle physics, matter science, the understanding of universal phenomena, and the understanding of the important scientific role neutrino particles play, leading to the discovery of lights, etc., to be further explained below.

B. Matter Having Atomic and Molecular Structures and Universal Phenomena

The matter having atomic and molecular structures made of electrons and protons also have high content of neutrino particles and they can collide interacting with atmospheric energetic neutrino particles to continue to produce the charge energy their protons and electrons needed to form it, to keep it stable, and to produce universal phenomena.

Upon entering Earth, the atmospheric energetic neutrino particles continue to lose their kinetic energy from their collision interactions with the particles of the matters of the Earth and, thus, come out of Earth having lower kinetic energy. The difference in the kinetic energy between entering and outgoing neutrino particles results in producing a neutrino-particles wind constantly blowing into Earth, which is the gravitational force of the Earth. Although their collision interactions are so ineffective that they are essentially undetectable experimentally in small scales, they are detectable in macro-scales. For example, the presence of gravitational force around heavenly bodies is detectable but it is undetectable around large buildings and even mountains.

Since the charge energy of both free electrons and protons are easily detectable experimentally, the efficiency of their collision interactions with the atmospheric neutrino particles should be high. Yet most matter having atomic and molecular structures essentially do not have detectable charge energy around them. It means that most charge energy of its protons and electrons has been used internally to form atoms and molecules, thus, turned into structural energy. Yet there are some compounds such as magnets having their atoms and/or molecules orderly arrange to have polarized charge energy detectable as magnetic force, which should be produced by having a neutrino wind passing through the compound. Also, the movement of electrons, which carry charge energy, to produce electric current or detectable neutrino-wind force surrounding the electric wire. Large amounts of charged particles such as clouds also create neutrino winds, which may play an important role in producing the destructive forces of Toledo and hurricanes. The above findings are scientific evidence to show that the collision interactions of matters with atmospheric energetic neutrino particles result in having atmospheric neutrino particles to produce universal forces. Also, the presence of an atmosphere of energetic neutrino particles produces both the so-called weak nuclear force making some atomic nuclei radioactive or unstable, producing gravitational and other universal forces, and the galaxy winds, which are the dominate force to shape up galaxies and by controlling the motions of the stars inside galaxies.

The scientific community has asked that If matter having atomic and molecular structures constantly collide interacting with the atmospheric energetic neutrino particles, these collisions should constantly produce heat and where is the heat? The answer is yes, they do. They produce such universal phenomena as that the center of planets are always hot and that there are volcanic activities on planets.

C. Light Particles

To be detectable experimentally, lights must be matter particles. Yet, having the fastest moving speed and essentially undetectably small mass, they cannot be the matter made of electrons and protons. Findings have shown that only matters such as those having atomic and molecular structures can emit and/or absorb lights and now we know that matters made of protons and electrons containing dense clouds of neutron particles. Findings also show that lights emitted by matter are formed by neutrino particles inside the dense clouds of the orbital electrons of matters such as that having atomic and molecular structures.

Neutrino particles have amazingly dense mass and a very broad particle-size distribution and their own collision interactions continue to reduce their average particle size. It is possible that inside the dense clouds of neutrino particles surrounding electrons inside matters some large single neutrino particles have also been stabilized from breaking up by collisions like both proton and electron by surrounding with a dense cloud of fine neutrino particles to carry mini-charge energy and to form mini-atoms and mini-molecules from different sizes of such stabilized composite particles, which are light particles. Since findings show that only matters, such as that having atomic and molecular structures, can emit and absorb lights, lights can only be made inside them. It means that only in the special environment of "inside the dense clouds of neutrino particles of the electrons inside matters under the constant bombardments of the atmospheric energetic neutrino particles, the mini-atoms and mini-molecules of neutrino particles or light particles are formed among those having composite-particles structure of different sizes." The formation of light matter particles inside the matters such as that having atomic and molecular structures is proven by both that everything has its own unique distinguishable color and that the collision interactions of its matter particles result in physical, chemical, and nuclear reactions emitting and/or absorbing light particles. Also, the light spectra of matter are quantized proving that lights are matter particles not massless waves.

The formation of lights by neutrino particles inside the clouds of neutrino particles surrounding electrons inside matters should have changed the physical and chemical properties of these dense clouds of neutrino particles. For example, unlike free protons and electrons, most of matter having atomic and molecular structures do not show charge energy around them. Besides, atoms and molecules are formed by protons and electrons with physical, chemical, and nuclear bonds, which have to be performed by the dense clouds of neutrino particles of orbital electrons showing that they have liquid-like properties to restrict their volume, shape, and to produce bonding forces.

As described earlier that charge energy has two opposite forces to express it. When the dense clouds of neutrino particles surrounding the electrons and the protons inside the matters have changed from gas phase to liquid-like phase, they have essentially lost their out-pushing force or their charge energy. It means that their charge energy has turned into the structural energy of matters.

Lights inside and outside matter play very important scientific roles. The lights emitted by matters have already been studies for several centuries but resulted in having more misunderstanding and confusions than real scientific understanding of them. Although the presence of light particles inside matters still has not been considered, they have always been included in scientific R&D, such as the study of the physical, chemical, and nuclear bonding. The new knowledge of light particles forming inside matters will certainly enhance future scientific R&D achievements. For example, it suggests that the orbital electrons of different compounds such as those of different metals are really different from one another due to having different light particles in them. With Tsau's new knowledge of lights and centuries of accumulated huge amounts of experimental findings of light R&D, it is time to advance the scientific knowledge of lights both inside and outside matters.

D. Cosmic Microwave Background (CMB)

The discovery of CMB in 1964 has been hailed to be the important experimental proof of the big bang theory which has predicted the presence of residue radiation from the big bang. Yet the overall findings have concluded that only matters emit lights and there is no scientific reason to suggest that CMB is not emitted by matter. The CMB findings should be interpreted by that CMB lights are emitted by matter having atomic and molecular structure inside an environment at very low temperatures surrounding our galaxy. It is apparent that CMB is emitted by the matter having atomic and molecular structure inside the galactic space surrounding our galaxy. The smooth distribution of CMB lights indicating that at very low temperatures, the crystalline structures of large heavenly bodies have been shattered and they have been broken up by random collisions.

E. Elementary Matter Particles—the One-Matter Nature Science

The most important conclusions from the R&D studies using particle accelerators are that proton and electron are the only two stable elementary particles carrying charge energy to collide interacting with each other to form unlimited numbers of stable matters (or matter objects) found in nature such as that having atomic and molecular structures. Tsau shows that both proton and electron are stabilized from breaking up by collisions because that they have a structure of composite-particles of a single large particle surrounded by a dense cloud of fine particles. Since proton and electron are the two stabilized elementary particle entities made of different sizes of the same elementary matter to make stable matters (matter objects) found in nature, it is a one-matter nature. This is true also for having light matter particles. They are also made from some stabilized entities having the composite-particles structure of a large neutrino particles surrounded by a dense clouds of fine neutrino particles.

Therefore, the nature has unlimited space filled with the particles of one kind of matter. Having amazingly dense mass these matter particles continue to be broken up by their collisions to finer particles. However, there are a few particles have been stabilized from breaking up by collisions by having a composite-particles structure of a large particle surrounded by dense cloud of fine particles and these stabilized entities, such as proton and electron, carry charge energies to collide with each other to form unlimited numbers of stable matters (matter objects) also exist in nature.

As the elementary particles continue to collide to break up to smaller sizes, there should be a high concentration of very fine elementary or neutrino particles taking up all space. It means that all matters (matter objects) are inside a medium of concentrated fine neutrino particles and there is no real vacuum space. Judging from the very low efficiency of collision interactions with matter having atomic and molecular structures, it is likely that only the larger particle sizes portion of the atmospheric neutrino particles plays the major roles in producing such universal phenomena as gravitational force, while a large portion of the finer sizes of the atmospheric neutrino particles are too small to contribute and it is likely that they also take up the space inside matter having atomic and molecular structures. But this may not be true for matters having higher mass density. Having such a medium in space can explain the universal redshift of starlight phenomena, why spaceships are slow down gradually in space, and more to be given in the scientific topics below.

F. New Cosmology and Astronomy

The acceptance of modern physics to be science by the scientific community has taken away the chance for the advancement of the mechanical physics-based science particularly in the past century; therefore still no mechanical physics based or the definition of science obeying cosmology and astronomy have been discovered. The definition of science demands or predicts that for matters made of protons and electrons with their charge energy such as matter having atomic and molecular structures and their heavenly bodies to exist, there has to be an atmosphere of tiny energetic matter particles in space to collide interacting with them to continue to supply charge energy to their protons and electrons or their structural energy for them to continue to form and be stable. Although an atmosphere of energetic neutrino particles has been discovered nearly a century ago to prove the prediction, the scientific community continues to insist that they penetrate through matters such as heavenly bodies freely and therefore have nothing to do with their existence and producing universal phenomena. As a result, the important scientific roles of the atmospheric neutrino particles have been denied by the scientific community for nearly a century already despite mounting experimental findings to show the contrary.

Today's cosmology accepted by the scientific community is the big bang theory based on Hubble's law using Doppler effect to interpret the phenomenon of universal redshift of starlight and the postulations that light is massless electromagnetic wave. It teaches that the universe started from a big bang or explosion of an extremely small and hot "singularity" and continues to expand. Although the theory defines what universe is, the outside universe is undefinable. Such a cooling universe should have no way to produce its own second-generation stars and should have died long ago when its first-generation stars somehow formed by the big bang of singularity had died. It is more religion-like belief than being science that the universe and its heavenly bodies can be produced by the big bang of a singularity with its existence in nature unexplainable. Besides, there is another possible scientific interpretation of the universal redshift of starlight phenomenon by having an atmosphere of tiny particles in space to collide interreacting with starlight, which are matter particles not massless waves, has not been considered. Since the theories of modern physics disobey the definition of science and are not science, the mechanical physics–based science should have its own correct scientific astronomy and cosmology waiting to be discovered.

1. The New Definition of the Universe

When temperature reduces, ideal gaseous particles move slower and, under constant pressure, take up less volume. When temperature approaches absolute zero degree, the theoretical gaseous volume approaches zero meaning that theoretically their random motions are stopped. Also, R&D findings at very close to absolute zero temperature show that atoms start to exhibit bizarre behaviors, such as that they condense, match in lockstep instead of fitting around independently to be called Bose-Einstein compensate matter. Unfortunately, temperature cannot be further dropped to absolute zero and below degrees to experimentally find out what will happen.

Having essentially stopped motion and vibration may also mean that the orbital electrons inside an atom can no longer circulate around atomic nuclei resulting in falling into them meaning that atomic and molecular structures are no longer stable at absolute zero and below degree temperatures. Tsau therefore postulates that zero-degree absolute temperature is the transition-point for matter having atomic and molecular structures to degrade by collapsing their orbital electrons into atomic nuclei to form a new matter called extremely dense matter (EDM) by Tsau having protons, electrons, and neutrino particles densely packed together, which is stable at all sizes only at zero-and-below degrees absolute temperatures. Since for the matter having atomic and molecular structures such as everything on Earth to exist in the universe, the temperature of the universe cannot drop to absolute zero and below temperatures. Tsau therefore proposes to define a universe to be an open space hosting matters, such as that having atomic and molecular structures, having sufficient number of stars in it to keep it above zero-degrees absolute temperatures.

2. The New Definition of the Outside Universe and *EDM*

Logically, there also should be place where does not have enough stars or have no star to keep its absolute temperatures above zero degrees. And Tsau further defines an open space hosting matters having absolute temperatures at zero and below degrees to be outside universe. Upon entering outside universe from a universe, heavenly bodies such as dead stars and planets will degrade to form EDM matter. It is expected that there is still a weak atmospheric pressure of neutrino particles, which gets weaker with farther away from a universe. Yet, having much denser mass than matter having atomic and molecular structures, EDM should still have strong gravity for them to be able to quickly grow their sizes to form EDM heavenly bodies. Having open space, EDM heavenly bodies in the outside universe also from time to time enter the universe.

3. The First-Generation Stars

The above postulations and definitions need experimental findings to prove their scientific correctness and values. Inside the universe, the nuclear reactions of all stars produce an atmosphere of energetic neutrino particles, proton, electron, hydrogen, and helium molecules. As shown earlier that the atmospheric energetic neutrino particles have very low efficiency of collision interactions with the matter having atomic and molecular structures but is still sufficient to produce the matter, to keep it stable, and to produce all universal phenomena.

The next scientific questions are where all the stars and their huge energy come from? Current teachings are that the large regular bright stars are formed by the collapsing of the hot molecular clouds of hydrogen and the death of these regular bright stars results in producing stars having extremely dense mass such as neutron stars, pulsars, and black holes. If the universe can continue to make its own stars, it can exist forever; however, many scientific questions remain unanswered, such as in the "constant expanding and cooling" universe how can the hot clouds of the hydrogen continue to form stars?

Since a universe relies on the energy of stars to exist, it should have no way to produce its own stars. The only possibility is that its new stars come from outside universe, where hosts EDM heavenly bodies. Upon entering the universe can EDM heavenly bodies turn into stars?

Upon entering the universe, EDM heavenly bodies will be bombarded by the atmospheric energetic neutrino particles. Having extremely dense mass, the efficiency of their collision interactions is expected to be very high resulting in producing huge amount of heat. Therefore we would expect that under the constant bombardments of the atmospheric energetic neutrino particles, the EDM heavenly bodies entered the universe will become so hot that they will start nuclear reactions. Based on our knowledge, nuclear reactions are chain reactions leading to violent explosions meaning that these EDM heavenly bodies should violently explode and be short lived. Yet we would expect that these EDM heavenly bodies entered the universe to become the densest stars like neutron, pulsar, and black hole stars but findings show that these stars have long lifespans.

It is possible that the nuclear reactions occurring there are not the kind of chain nuclear reactions we know and they may be some kind of sustained nuclear reactions unknown to us. One possibility is that nuclear reactions can only occur on the surface of EDM heavenly bodies where the bombardments of the atmospheric energetic neutrino particles are the strongest. The new astronomy therefore teaches that neutron, pulsar, and black holes stars are the first-generation stars and they are the 100 percent EDM heavenly bodies entered the universe from outside universe. Having amazingly dense mass and long-lasting lifespans may mean that they are not penetrable by a portion of the larger-sizes atmospheric energetic neutrino particles and their collision interactions on their surface result in producing sustained fusion reactions occurring only on their surface resulting in emitting hydrogen, helium, and other elementary matter particles for them to become the first-generation stars.

These first-generation stars constantly undergoing nuclear reactions on their surface to emit energetic particles should have very strong antigravitational force and findings show that they do. For example, they throw hydrogen and helium far away out and some other evidence will be given later. Also, if a large portion of the atmospheric finer neutrino particles can penetrate through and collide interacting with these stars, they should have strong gravitational force and that there are findings having also confirmed so. Therefore these densest stars unexpectedly have both strong gravity and strong antigravity. The science of both antigravity and gravity will be further explained and discussed later. These densest stars are essentially invisible since they are not the matter having atomic and molecular structures emitting lights.

The nuclear reactions on the surface of these densest stars are due to constant bombardments of atmospheric energetic neutrino particles on their surface. Therefore the starting materials of their fusion reactions are electron, proton, and neutrino particles while hydrogen and helium are reaction products. Using hydrogen molecules as fusion-reaction starting material to form helium is not exactly the kind of fusion reactions occurring in stars and may be more difficult to achieve.

4. The Second-Generation Stars

A densest star constantly undergoes nuclear reaction on its surface to produce hydrogen and helium gas molecules throwing them to a faraway place, where they gradually gather to form a large hollow ball or donna shaped outer-layer, which gets denser and brighter with time, due to that the nuclear reactions of the densest star at its center continue to produce hydrogen and helium and to heat it, to gradually form a regular large bright star or bright donna-shaped rings surrounding a black hole. Pulsar stars and many black holes are fast rotating concentrating emitting their nuclear-reaction particles or antigravity on a thin plane at an angle in their rotation direction to also push their outer-layer to rotate. Therefore the second-generation stars are the first-generation stars surrounded by an outer layer of matter having atomic and molecular structures.

As expected, the inner surface of the outer-layer of a mature regular bright star will reach the critical temperatures to start nuclear reactions to form elements heavier than helium making the outer-layer harder and harder for the gases produced by the nuclear reactions of the neutron or pulsar star at its center to dissolve in and to penetrate through, thus, building up the inner gas pressure to eventually lead to a violent expansion or supernova explosion of the outer-layer turning a regular bright star to a red giant star, exploding to shed part of its outer-layer to become a white dwarf, or to shed all of its outer-layer to reborn a smaller neutron or pulsar star. In summary, the first-generation stars are pure EDM heavenly bodies entered the universe from outside universe and they begin to produce the second-generation stars having an outer layer of matter having atomic and molecular structures. The second-generation stars are the black holes having a bright donna shaped outer-layer, regular bright stars, the red giant stars, the white dwarf, and the brown dwarf stars.

Some planets, such as Earth, are made of the remnants of the supernova explosions of regular bright stars, red giants, or white dwarfs. The other type of planets are giant gaseous planets mainly containing hydrogen and helium, which are formed from cooling off brown dwarfs due to that the densest stars at their center get too small to continue their nuclear reactions.

The presence of the second-generation stars and their supernovae explosions further demonstrate that the densest stars have strong antigravity.

5. Galaxy Wind of Neutrino Particles

A galaxy has billions of stars to produce a galaxy neutrino wind blowing out of the galaxy and when all galaxy winds meet, they create a hurricane-like neutrino wind to push all the stars of a galaxy moving like the clouds inside a hurricane known to be the spiral galaxies. Since most stars inside a galaxy are far apart from one another, the galaxy wind is the dominant pushing-force controlling the motions of its stars not the gravitational force among stars as taught by the scientific community.

6. Science of Gravity and Antigravity

The very low efficiency collision interactions between the atmospheric neutrino particles and heavenly bodies having atomic and molecular structures result in producing gravitational force, which is the neutrino-particles wind constantly blowing into them. The atmosphere of neutrino particles is also the medium of all matters. Due to their constant collisions with each other to break themselves up to finer particles, the medium should have a large portion of very fine particles. Having very dense mass, the densest stars have much higher efficiency of collision interactions with the atmospheric neutrino particles. Findings show that the efficiency of their collisions is so high that their surface undergoes nuclear reactions to constantly emit elementary matter particles, hydrogen, and helium particles resulting in having strong antigravity. Yet large amounts of findings also show that they have strong gravity. The question is why their antigravity does not cancel their gravity?

The above findings suggest that there is more than one mechanism for the collision interactions of the atmospheric neutrino particles with the two types of matters having very different mass density to produce gravity and antigravity. We would expect that only the portion of the larger neutrino particles is mainly responsible for their very low efficiency collision interactions with matter having atomic and molecular structures to produce their gravity while the smaller particle sizes portion may be able to pass through them essentially freely, thus, having only little contribution to produce its gravity.

Since the EDM heavenly bodies or the densest stars have very dense mass, the larger particle sizes portion of the atmospheric neutrino particles cannot penetrate into them and their bombardments on surface leading to undergoing nuclear reactions there for them to have strong antigravity. Also, the portion of the smaller particle sizes of the atmospheric neutrino particles may be able to penetrate through these densest stars to collide interacting with them to produce their strong gravity. Findings do show that the densest stars have both strong antigravity and strong gravity, such as that some bright stars orbit around each other or around something invisible like densest stars. Yet their strong gravity does not seem to cancel their strong antigravity. For example, densest stars inside bright stars do not suck their outer layer in. The reason may be, as discussed earlier, the strong gravity of the densest stars is produced by the portion of the finer particle sizes of atmospheric neutrino particles, which has little contribution to produce the gravity of the matter having atomic and molecular structures. Therefore the strong gravity of the densest stars mainly acts on themselves and on the densest stars inside the second-generation stars but not effectively on matter having atomic and molecular structures such as their outer layers and planets.

7. Stars Having Planets

The strong antigravity of the densest stars is also shown from the following findings. A pulsar fast rotates emitting most of its nuclear-reaction particles or concentrating its antigravity on a thin plane at an angle along its rotating direction to push the outer-layer of its regular bright star to rotate in the same direction the pulsar rotates. Besides, its concentrated antigravity can support its bright star to have planets having pseudo-steady state orbitals only in the rotating direction of the pulsar. Since only the pulsar stars' antigravity can be used to support their planets to have pseudo-steady state orbital motion, only the bright stars having a pulsar at their center or rotating outer-layer can have planets such as our Sun. Many black holes are also fast rotating, thus, push their donna-shaped outer-layer to rotate.

8. Possible New Kind of Matter Having Strongest Gravity

Findings show that in space there are stars orbiting around something invisible having very strong gravitational force. They are suspected to be black holes but black holes usually have bright donna-shaped ring surrounding them. The findings suggest that there may be a new kind of matter invisible having very strong gravity.

It is possible that some EDM heavenly bodies entering the universe collide with heavenly bodies such as bright stars and planets to be coated with a layer of matter having atomic and molecular structures thick enough to protect their surface from direct bombardment of atmospheric energetic neutrino particles to start and to undergo nuclear reactions. This kind of matter is expected to have very strong gravity and it may be invisible since the light emitted by its coating matter can be destroyed by its strong gravity wind. It may also be due to that this new matter is too far away and too small to be visible.

APPENDIX
The Story of Tsau's Scientific Revolution

Tsau is a PhD (McGill University, 1970) physical chemist having a R&D career in pharmaceutical companies retired early in 1996 at age sixty. Having a classical mechanical-physics background, he has difficult to understand the newly introduced modern physics. His early retirement gave him time to study modern physics to unexpectedly lead him to dedicate his retirement to prove that modern physics is not science, to advance classical mechanical physics, and to start a scientific revolution to save science.

His efforts to understand the theories of modern physics have come up with a surprising and unfortunate conclusion that modern physics using mathematical-derivation method to make discovery and to interpret universal phenomena is contradictory with the well-established mechanical physics using experimental method to do so meaning that modern physics cannot be correct science. He began to search for the scientific answers to modern physics or advancing mechanical physics–based science. For example, he has been able to use the experimentally-proven to exist pushing force taught by the mechanical physics to replace the attraction forces acting from distance taught by modern physics. He has soon found that his findings based on mechanical physics are contradictory with the teachings of modern physics, thus, are not acceptable for publication in scientific journals. He has begun to seek support from National Science Foundation (NSF) and to write to such important scientific organizations as American Institute of Physics (AIP) and the board of education of many states without being able to catch their attention. Still, he has continued his own effort to advance mechanical physics–based science and to prove that modern physics is not science.

It had been twenty-six years since his retirement and he has finally reached the overall scientific conclusions with which he is happy. He has rediscover that science has a specific definition or scientific method, the experimental method, which must be obeyed to be science. Developed based on postulations disobeying the definition of science and using mathematical-derivation method not experimental method, the modern physics developed and taught as science by the scientific community is not provable to be correct science experimentally or not science. Besides, Tsau's breakthrough scientific discovery to show that mechanical physics–based science now can interpret everything detectable experimentally meaning that it alone is the entire correct science excluding modern physics.

Science is too important to humankind and the corruption of the scientific community including governmental authorities to develop and to teach modern physics as science has already done unimaginable harm. Even though over the past three decades Tsau has continued to ask scientific authorities to correct the existing serious scientific mistakes, to learn, and to evaluate his breakthrough scientific discoveries, his requests have been entirely ignored. Recently, he has started a new round of efforts to save science and science education as given below.

A. Legal Challenge of Scientific Correctness

In realizing that US laws are based on the science of mechanical physics alone to protect human right, Tsau has tried to legally save science. He has sued both NSF and AIP (American Institute of Physics) in Chicago federal court more than once but having all lawsuits been dismissed without getting a chance to discuss science in court or convincing these scientific organizations to face serious scientific mistakes. His latest lawsuit was filed against the federal government on February 18, 2020. Both the lawsuit and the judge's reasons of dismissing it for lack of subject matter jurisdiction are given below. All his lawsuits are pro Se cases and his poor knowledge of law may play an important role in the failure of his legal challenges.

1. The Most Recent Lawsuit

United State District Court

Northern District of Illinois Eastern Division

Josef Tsau	)	Case No. 1:20-cv-01021
Plaintiff	)	
v.	)	Judge: Sharon Johnson Coleman
US Government	)	
Defendant	)	

Complaint

Plaintiff alleges that our government has long been corrupted to turn science into a religion violating the law of The First Amendment of Constitution. As a US citizen, he is obligated to help saving our science, science education, and to protect our law to file this complaint. To overcome government's immunity from lawsuit, he has written letter to Attorney General William Barr on August 6, 2019, to request his investigation of the corruption without response. Besides, also in 2019 plaintiff has written letters to the department of education or board of education of more than twenty states to report the corruption in science education without having a response from them to address the issues. Plaintiff has exhausted administrative means.

Science has been discovered by Copernicus and Galileo during the sixteenth and seventeenth centuries and is also known to be the mechanical physics–based science, which has led to both industrial revolution and today's civilization. It has a specific definition of using experimental methods to study, to discover, and to prove nature's laws allowing everything to have unique experimental findings–based interpretation and all interpretations being consistently coherently and logically related as a whole. Since experimental methods can only detect or study matter, the scope of science is limited by its definition to study matter and the phenomena produced by matter. To separate science from religion, the Supreme Court has relied on the definition of *science* to rule that both creationism and intelligent design are religious teachings not science since they are not provable correct experimentally. Also, to study matter science has a clear secular purpose. This feature of science has also been utilized by law to apply the three-part Establishment Clause test for the US government set forth in Lemon v. Kurtzman, 403 US 602 (1971). Under that test, to satisfy the Establishment Clause, the US government must (1) reflects a clear secular purpose; (2) has a primary effect that neither advances nor inhibits religion; and (3) avoids excessive government entanglement with religion, Committee for Public Ed. and Religious Liberty v. Nyquist, 413 US 756 (1973). Therefore both science's specific definition and feature of having a clear secular purpose have been legally applied to differentiate science from religion to protect the law of the First Amendment of the Constitution.

Plaintiff alleges that the academic scientific community has long been corrupted to lead to the corruption of the scientific community, including the governments worldwide, to turn science to a religion. It began in the seventeenth century from its accepting Newton's theory of gravitation assuming gravitational force to be the attraction force among heavenly bodies acting from distance without having matter particle in space to produce it, thus, violating the definition of *science* that science studies matter and matter particles only. Later, in the twentieth century, this theory has essentially been replaced by Einstein's theory of general relativity teaching that gravitational force is curved space-time again having no matter particle in space to produce it. During the nineteenth century, Maxwell found that his mathematically derived electromagnetic-wave equation had a theoretical speed magically matched that of light found experimentally and he therefore proposed that light was electromagnetic wave. During the twentieth century, Einstein's theories of relativity had accepted his proposal and further postulated based on the electromagnetic-wave equation that light had no mass and an absolute speed and also that mathematical-derivation had been proven experimentally to be a valid method to make scientific discoveries. Upon accepting the above theories by the scientific community, light has become wave, which without having mass and relative motion, cannot be matter particle. Yet any violation of the definition of *science* results in being outside of the scope of science, not science or a religion according to Supreme Court's ruling. For example, a wave or a force not produced by matter particle in space should be undetectable or unprovable to exist or to be anything experimentally and therefore curved space-time or electromagnetic wave has not been proven to be light or gravitational force and mathematical derivation has not been proven to be a valid method to make scientific discoveries. Besides, since both the real light and the real gravitational force are experimentally detectable, they have already been proven to be matter particles experimentally. Also, in the twentieth century, both the quantum theory and the big bang theory have been developed based on the same postulations of Einstein's theories of relativity including using mathematical derivation to make scientific discoveries and the above theories have been combined to be today's mainstream-of-thought physics long been accepted as proven-correct science by the scientific community and the governments worldwide. Plaintiff alleges that since mainstream-of-thought physics is based on postulations violating the definition of *science*, it is not science but a religion. For example, its discovered universal phenomena by mathematical-derivation method, such as its discovered gravitational force and light, are not produced by matter particles, thus, are not detectable or provable experimentally to exist. It therefore is a religion teaching a fantasy universe with its universal phenomena not really provable to exist experimentally.

Plaintiff alleges that our scientific community including our government has long been using the mainstream-of-thought physics, a religion, to interpret such important scientific topics as the universe, the universal phenomena, the matter, the particle physics, the astronomy, and the cosmology by denying and blocking the use of (the mechanical-physics based) science to interpret them despite that experimental finding continuously accumulated support science only. For example, the scientific community has discovered an atmosphere of energetic neutrino particles, produced by the nuclear reactions of all stars, in space surrounding all matter nearly a century ago. It is known that matter having atomic and molecular structures is made of electrons and protons relies on their charge energy and the atmospheric energetic neutrino particles happen to be the only possible supply of the charge energy. In order to protect its religious teachings, such as that its universal phenomena do not need matter particle to produce them, the scientific community, including our government, has long been teaching and continue to insist that the matter is essentially empty space, and neutrino particles are so inactive that they freely pass through matter without having collision interaction with it to produce the matter and universal phenomena despite that now the accumulative experimental findings overwhelmingly prove the contrary.

The centuries-long accumulated experimental findings, however, have led plaintiff to discover the advanced (mechanical physics–based) science and to further prove that science can interpret all the compositions of matter and phenomena by itself to be the only correct science, as given in Plaintiff's book *The Beautiful Mechanical Universe*, published in 2014 by Lulu. It further shows that science interprets everything differently from those of the mainstream-of-thought physics already accepted worldwide as proven correct science, thus, having conclusively proven that mainstream-of-thought physics is not science, a religion, or obsolete. Therefore science teaches an entirely different scientific world from what we have learned and what students are learning in schools and colleges. Briefly, science's gravitational force is the wind or the pushing force of the atmospheric neutrino particles blowing into heavenly bodies. Its light and heat are the mini-atoms and mini-molecules made of neutrino particles inside matter. Matter also contains neutrino particles to become dense enough to collide interacting efficient enough with the atmospheric energetic neutrino particles to form different compositions of matter, to produce universal phenomena, and even to undergo nuclear reactions. Its elementary particles to make matter of different compositions are all the particles constantly produced and emitted by the nuclear reactions of stars divided into the four particle-size groups of the (noncharged) single particle of protons, the (noncharged) single particle of electrons, the neutrino particles, and the mini-neutrino particles. A single particle of proton or electron always has a dense cloud of neutrino and mini-neutrino particles surrounding it to be efficient enough to collide interacting with the atmospheric energetic neutrino particles to gain their kinetic energy converting it to charge energy. Also, a universe is defined by the space having enough stars to keep its temperature above absolute zero degree, which is the transition point for the matter having atomic and molecular structures to collapse to form a very dense matter (VDM) having protons, electrons, neutrino, and mini-neutrino particles densely packed together. The VDM heavenly bodies are stable at any size only at absolute zero and below temperatures outside the universe, where they form and grow in size by collecting surrounding VDM and other elementary particles. Once entering the universe, a VDM heavenly body is constantly bombarded by the atmospheric energetic neutrino and mini-neutrino particles and since VDM is so dense that even neutrino and mini-neutrino particles cannot penetrate deep into it. Therefore inside the universe the collision interactions between a heavenly body of VDM and the atmospheric energetic neutrino and mini-neutrino particles are so violent that its surface starts and continues to undergo nuclear reactions to emit hydrogen, helium, and other elementary particles to become the densest stars known as neutron stars, pulsar stars, and black holes and that they all should have strong antigravity not strong gravity as taught in schools. All other stars such as regular bright stars, red giant stars, white dwarfs, and brown dwarfs are made having a VDM at center and an out-layer of matter having atomic and molecular structures.

Having a scope covering all the compositions of matter and all phenomena detectable experimentally, supported by all accumulated experimental findings, and strictly obeying the definition of *science*, the (mechanical physics–based) science now has conclusively proven itself to be the only correct or real science. Yet the advance science discovered by plaintiff is unacceptable both for publishing by the scientific community and for funding by our government. Plaintiff alleges that although it is the responsibility of the scientific community and our government to support the R and D and the teaching of the only real science, they have long been corrupted to develop and to teach a religion in the name of science while acting and teaching against the only real science.

In summary, plaintiff alleges that our government has long been corrupted to financially support the R&D and the teaching of a religion known as the mainstream-of-thought physics in the name of science such as through NSF's and DOE's funding of its teaching in schools and its R and D. The science-teaching standards for schools our government such as its NSF and DOE having helped, guided, and financed the states to write up and to publish known as the Next Generation Science Standards (NGSS) has officially abandoned the definition of *science* by adding mathematical derivation as a method for make scientific discoveries to turn science education into a religious teaching. Schools have long been teaching the religion as science such as its gravity and light theories. The teaching and the development of a mathematically derived fantasy universe having its universal phenomena not detectable or provable experimentally to exist do not have a clear secular purpose to meet the above-mentioned test requiring for US government and also fit the rulings of the Supreme Court to be a religion. Plaintiff further alleges that our government has both a primary effect in the developing, advancing, and in being extensively entangled with the religion by financing its R&D and its teachings in schools as science and in using its knowledge to establish governmental science policy.

Plaintiff respectfully asks the court to stop the corruption of our government to save our science, science education, and to protect the law of The First Amendment of Constitution by ruling that science has a specific definition protected by law and any teaching and theory violating the definition of *science* is a religion not science and that the mainstream-of-thought physics is a religion not science.

Undoubtedly, our government itself is responsible and dedicated to supporting the R&D and the teaching of the only real science in schools. Hopefully the lawsuit can convince it to understand that the mechanical physics–based science is the only real science and to take on its responsibility to solve our current existing scientific problems. If so, plaintiff will be happy to withdraw the lawsuit.

Respectfully submitted, Josef Tsau.

2. Court's Orders

The Order: the lawsuit has been dismissed for lack of subject matter jurisdiction.

Background: Construing pro se plaintiff Josef Tsau liberally, *Chronis v. United States*, 932 F 3d544, 554 (7th Cir. 2019), he alleges that "our government has long been corrupted to turn science into religion in violation of the law of the First Amendment." Tsau takes issue with the National Science Foundation and the Department of Education "science teaching standards" that allow schools to teach religion as science. He argues that the government is responsible for the teaching of "real science" in the public schools.

This is not the first time that Tsau has brought a lawsuit in this district challenging the government's policies on science and science education. See, e.g., Tsau v. National Science Foundation., Nos 17-cv-3966; 1o-cv-6323; 04 C 5634. In those cases, the courts dismissed Tsau's claims for lack of subject matter jurisdiction.

Discussion: Federal courts are courts of limited jurisdiction and are obligated to consider subject matter jurisdiction at any stage of the proceedings. (citation omitted) "Article III extends the judicial power only to the resolution of cases and controversies," and "requires a plaintiff to have suffered an injury-in-fact traceable to the defendant and capable of being redressed through a favorable judicial ruling," also known as standing. (Citation omitted) "To establish injury in fact, a plaintiff must show that he or she suffered [an invasion of a legally protected interest] that is [concrete and particularized and [actual or imminent, not conjectural or hypothetical.]" (Citation omitted)

Tsau asserts that as a United States citizen, he is obligated to help save science and science education. The remedy he seeks is for the court to "stop the corruption of our government to save our science, science education, and to protect the law of the First Amendment." Missing from his complaint, however, is Tsau's own First Amendment injury traceable to the government's conduct that this court can remedy. Although the education of the nation's is imperative, without more, Tsau cannot establish Article III standing that would allow him to proceed with this lawsuit.

Last, the court notice that "the payment of taxes is generally not enough to establish standing to challenge an action taken by the federal government. In light of the size of the federal budget, it is a complete fiction to argue that an unconstitutional federal expenditure causes an individual federal taxpayer any measurable economic harm." (Citation omitted) The court therefore dismisses this lawsuit for lack of subject matter jurisdiction. IT IS SO ORDERED.

3. I have filed an Appeal

Again, it has been dismissed by appeal court with warning not to continue the lawsuit. My legal challenge had been stopped. Yet I continue to believe that science needs and deserves legal protection and wonder whether having a lawyer can change the outcome.

B. Efforts To Convince Scientific Authorities

I have been encouraged by that our newly elected President Biden have raised the importance of science to Cabinet level and have written to him hoping that he, with his science team, can lead the nation and the world to save and to advance our science, again, resulting in nonresponse. He then has started his latest round of efforts to write to NSF, NASA, and Department of Energy. The letter to NSF is given below.

1. My Letter to NSF

August 25, 2021

Dr. Sethuraman Panchanathan
Director
National Science Foundation
2415 Eisenhower Avenue
Alexandra, Virginia 22314

Dear Dr. Panchanathan:

Since it is the responsibility of NSF to advance correct science and its applications, I am submitting to you, the head of NSF, the attached paper of mine entitled "Reinventing Both the Definition and the Science of Everything Detectable Experimentally." It summarizes the science obeying the definition of *science* including my breakthrough scientific discoveries to show that it can interpret everything detectable experimentally logically, consistently, and coherently related with one another by experimental findings as a whole to conclusively prove that it, the science obeying the definition of *science*, alone is the correct or real science. Yet it teaches away from today's accepted science mainly due to that the modern physics we have accepted as science for more than a century has now been proven wrong or not science by both that it disobeys the definition of *science* and that the definition-obeying science teaches everything different from modern physics.

My breakthrough discovery is a wonderful scientific world, having answered centuries-old scientific questions such as what are gravitational force, light, and universe with neutrino particles playing important roles, waiting for NSF's evaluation and advancement. In the fiercely competitive world of science, it should be a good opportunity to be ahead in a newly discovered scientific R&D direction.

We also need to reeducate the general public and to revise our science education essentially by removing modern physics from science. It has taken a (Einstein's) scientific revolution to develop and to make modern physics the most popular science, thus, making it difficult to do so. In my opinion, coordinated efforts from different scientific and educational organizations and a strong leadership are needed to accomplish the above scientific missions and you are in the position to be the important leader.

I have been trying for over two decades to convince scientific authorities that mechanical physics is largely the correct science but modern physics is mostly wrong or not science. I have even brought both NSF and American Institute of Physics to federal court in Chicago hoping to have a chance to debate on scientific issues in court without success. Working with the issues involving the entire scope of science by myself, I have to continue to improve my scientific comprehension or even correcting some scientific views and also have difficulties to clearly express them. I am finally happy with the paper submitting to you and hope that you could understand and forgive my past lawsuits against NSF for the purpose of saving science.

As usual, no response has obtained.

2. Letter to Congress

Hoping that congress may be able to pass a law to protect science, he has written the following letters to Congress copying President Biden, and NSF hoping that they may discuss his proposed scientific issues coming up with a way to save and to advance our science.

October 12, 2021

Ms. Eddie Bernice Johnson
Chairwoman
Committee on Science, Space, and Technology
2321 Rayburn HOB
Washington, DC 20515

Dear Ms. Johnson:

I allege that our scientific community including the scientific organizations of our government has long been teaching wrong science and making wrong scientific decisions. I further accuse that it has repeatedly ignored the breakthrough scientific discoveries of the correct or real science to cover up the existing serious scientific mistakes. It has been over two decades since I have started writing to important scientific organizations to point out that serious mistake exists in science and that mechanical physics–based science should be the only correct science but has been repeatedly ignored. I have even brought both American Institute of Physics and NSF to the federal court in Chicago more than once for my mission to save science but the cases have been dismissed without having a chance to debate science in court. It is troubling that even my repeated lawsuits cannot lead these scientific organizations to investigate existing scientific mistakes. Recently, I have again submitted the same scientific paper as the attached to NSF, NASA, Department of Energy, and President Biden requesting their scientific investigation and actions to save science and again have been completely ignored. Hopefully Congress will investigate the existing serious scientific mistakes and their cover up to save our science, science education, and to advance science.

May I also point out that although Congress has long recognized the importance of science to form such important governmental organizations as National Science Foundation (NSF) to fund scientific R&D and education, it has not legally defined what (nature) science is. According to my scientific discovery, the (nature) science discovered by both Copernicus and Galileo has a specific definition of "using experimental method as the specific scientific method to make scientific discoveries," which must be strictly obeyed for the science to have the unique features of having only one scientific answer for everything with all answers logically, consistently, and coherently related to one another by their experimental findings. Besides, since experimental method can only detect or study matter particles, which have both mass and relative motions, science has limited its scope to study matter particles only. The major problem of science since its discovery in the sixteenth and seventeenth centuries has been that the definition of *science*, including its scope limits, has not been officially recognized and as a result often not obeyed.

According to my findings, the mechanical physics–based science has largely obeyed the definition of *science* to be largely correct science. Yet the long historical hot scientific debates on both whether lights were matter particles or massless wave and whether there were tiny (matter) particles in space to produce gravitational force or not had directly put the definition of *science* on the debate line. Unfortunately, scientific community had finally chosen to accept both debate answers disobeying the definition of *science* leading to its developing and teaching the following (un)scientific theories disobeying the definition of *science*: Newton's gravity theory, Maxwell's theory of light, Einstein's theories of relativity, quantum theory, and big bang theory or today's modern physics. Modern physics uses mathematical-derivation method to make scientific discoveries, which have to be different from those made by experimental method and therefore cannot be science. For example, to obey the definition of *science* there must be an atmosphere of energetic tiny matter particles in space to collide interacting with the matter particles to produce gravitational force to be experimentally detectable but it is incompatible with both Newton's and Einstein's mathematical-derived gravity theories. Still, modern physics has long been the accepted correct science playing a very important role in today's science to interpret such important scientific topics as astronomy, universal phenomena, cosmology, particle physics, and even matter. Yet apparently it is not science teaching a fantasy universe having its universal phenomena such as its light, gravitational force, relativistic phenomena, space-time, etc., not really detectable or provable to exist experimentally. In summary, a long chaotic era of the scientific community has existed for centuries due to that the scientific community continues to develop and to teach both mechanical-physics and modern-physics based sciences contradictory with each other and one of them must be wrong science or not science.

If the scientific community were scientifically correct to develop and to teach modern physics as science, my breakthrough scientific discovery in mechanical physics can never happen. Now mechanical physics can interpret everything detectable experimentally including those already interpreted by modern physics accepted by the scientific community as the correct science but the experimental findings-based interpretations are entirely different from those of modern physics, again, proving that modern physics is not science. My scientific discovery has new definitions for both the universe and the outside universe and the universe is created by the collision interactions of its atmospheric energetic neutrino particles with matter. Their very low efficiency collision interactions with matter having atomic and molecular structures are sufficient enough to make the matter itself and to produce universal phenomena. Besides, their 100 percent effective collision interactions with the extremely dense matter (EDM) heavenly bodies, formed in the outside universe and entered the universe, result in undergoing unclear reactions on the surface of the EDM heavenly bodies to become the first-generation stars in the universe, which are neutron, pulsar, and black holes. Please read the enclosed science paper entitled "Reinventing the Definition and the Science of Everything Detectable Experimentally" for more details of my breakthrough scientific discovery.

Both the definition of *science* and my breakthrough scientific discoveries in mechanical physics–based science have conclusively disproved modern physics to be science. It is time for Congress to legally define what (the nature) science is by legally defining it with its definition to end the long chaotic era of the scientific community, to save, and to advance the correct or real science. It is to be noted that since the forensic science is the pure mechanical physics–based science obeying the definition of *science*, US courts have long been using experimental findings alone to protect people's fundamental right. Just imagine, if the mathematical-derivation method is also used to create a chaotic court system having two or even more legal answers.

Mathematics has always been a good tool for interpreting experimental findings, for example, it is useful to quantify force, energy, the efficiency of vaccines, etc., but it cannot be used to make scientific discoveries. Scientific community teaches that mathematical-derivation method has discovered the equation of electromagnetic wave, which has been proven to be light since its theoretical speed matches that of light found experimentally. Yet, according to its equation, electromagnetic wave has no mass and an absolute speed meaning that it is not matter particle and therefore it should not be detectable or provable to exist experimentally like the real light and, thus, it cannot be the real light. In another example, gravitational force can be accurately calculated by mathematical theories to predict planetary motions. Yet these theories, such as those of Newton and Einstein, do not have matter particles in space to produce their gravitational forces meaning that their gravitational forces are not detectable experimentally in space, but the real gravitational force is. Even Einstein's God Equation ($E = mc^2$) cannot be science. Although the explosion of atomic bomb appears to have lost some mass but not that of hydrogen bomb. Besides, Science teaches that only the collision interactions of matter particles in space can produce forces and energies but the existence of the energy converted by matter, which have no mass and no relative motion, is not science or not definable scientifically.

It looks like a case of one crazy scientist up against the scientific community worldwide but it is not. It is the two contradictory sciences, the mechanical-physics and the modern-physics based sciences the scientific community has developed and taught for several centuries as (one) science, up against each other. It is also the centuries-long legal question "what is science?" waiting for answer. In fact, most scientists and engineers in their professional carriers have only applied the mechanical physics–based science and technology and if they have a chance to learn my breakthrough scientific discovery, they will be on my side. Also, now it should be an easy scientific question to solve since the correct or the real science always stands tall supported by all experimental findings and there are huge amounts of accumulated experimental R and D findings available to support the correct or real science. In fact, science is easily and logically understandable even by the majority of the general public and difficult mathematics plays no role in understanding science.

I respectfully request congress to start an investigation process to determine which the mechanical-physics or the modern-physics base science is the correct or real science including evaluating my breakthrough scientific discovery in mechanical-physics based science and to investigate the damage of Having the existence of the centuries-long chaotic era in the scientific community. Again, the attached paper summarizes all my scientific discovery. Upon removing the influence of the modern physics, with the strong accumulative scientific evidence the paper has given the important scientific roles the atmospheric energetic neutrino particles deserve to discover a beautiful neutrino particle dominated scientific world. Hope that my breakthrough scientific discovery will soon be benefiting humankind. May I suggest forming a special scientific committee to do so. Hopefully the findings will convince Congress to legally define what science is by legalizing the definition of *science*, which has long been badly needed to end the long chaotic era in scientific community, to save, and to advance science.

Tsau then has written a follow-up letter:

October 23, 2021

Dear Ms. Johnson:

I would like to follow up my submitting scientific paper and letter to you on October 12, 2021, by writing to you and government again hoping to convince you all to talk and to work together to solve the centuries-long serious scientific mistakes made by our scientific community due to its unknowingly disobeyed the definition of *science*, which is still not been legally recognized to legally define what is "nature" science, and to properly handle my breakthrough scientific discovery to save and advance science and science education.

The definition of (*nature*) *science* specifies using experimental method as the specific method to study the nature and to make scientific discoveries and it has led to the developing and the teaching today's mechanical physics–based science. Yet our scientific community also has long been misled by the concepts disobeying the definition of *science* to develop and to teach the theories of modern physics using mathematical-derivation method to make scientific discoveries. Since the mathematically discovered universal phenomena and light do not need an atmosphere of tiny matter particles to produce them, the scientific community has been teaching that the atmospheric neutrino particles, found nearly a century ago, are so inactive that they do not collide interacting with matter. Yet the definition of *science* predicts or requires the existence of an atmosphere of energetic tiny matter particles such as the energetic neutrino particles in space to collide interacting with one another and with matter to produce both matter itself and its phenomena.

Scientific community teaches that mechanical physics is incapable of interpreting relativistic and quantum phenomena and modern physics is developed to do so. But since relativistic phenomena are not experimentally detectable or provable to exist, they are not science. Since all experimentally detectable phenomena including light should be produced by matter particles, they, such as the light spectra of matter, are naturally quantized.

With my breakthrough scientific discovery, now mechanical physics–based science teaches everything detectable experimentally logically consistently coherently related to one another with experimental findings. This feature alone has conclusively proved its scientific correctness. Besides, unlike the theories of modern physics, it is supported by tons of accumulated experimental findings. It is very unfair and perhaps even illegal to me and to the general public for our government to ignore and to cover my scientific discovery up.

A major scientific problem has been to abuse experimental method to prove modern physics to be science. For example, the expensive particle accelerators have been heavily relied on to do so. According to my discovery, both proton and electron have a composite structure of a large single (uncharged) proton or electron particle surrounded be a dense cloud of neutrino particles to collide interact with the atmospheric energetic neutrino particles to become charged. The charge energy relying on such a structure to produce is expected to be unstable to motion and acceleration. Therefore charged particles cannot be accelerated to light speed is not due to their mass increases with speed but due to losing charge energy. Without scientific understanding what is charge energy, the findings from particle acceleration have largely been misinterpreted and useless.

It is to be noted that since both the gravitational force and the light of modern physics are not produced by matter particles, they are undetectable experimentally but both real ones are. All experimental findings only support the experimentally detectable ones, which are the matter particles to produce them.

Without recognizing that the atmospheric energetic neutrino particles colliding interacting with matter to produce matter itself and universal phenomena, the scientific community does not accept that its wind is the dominant force shaping up a galaxy and controlling the movement of its stars. Postulating that the gravitational force of stars is the dominant force in a galaxy to control the motions of stars, modern physics has theorized that only about 5 percent of matter in the universe has been found while 95 percent of them are undetectable dark matter and dark energy. The religious nonscientific theories of modern physics can be easily created to waste governmental R and D budget, to block the progress of science, and to misled science education.

Again, the definition of science must be obeyed to be the only correct or real science. Hope that now our scientific community can voluntarily do so itself to correct its centuries-long serious scientific mistakes. Otherwise, I respectfully ask congress to legally define what "nature" science is to save, to advance our science, and our science education.

Cc:
President Joe Biden
Dr. Sethuraman Panchanathan, NSF
Again, no response has been received. Now he has exhausted his ways and ability to persuade scientific authorities to save and to advance science.

C. Time for the General Public to Be the Real Owner of Science

Hopefully this simple and logical to understand science book can show the general public that everybody can understand science and that the definition of *science* must be obeyed to be science for everybody to become the real owner of science. Once a good portion of the general public become the real owner science, they will be able to influence and even demand our scientific community including our government to develop and teach correct science only.

The mission of this book is also to lead a scientific revolution to correct centuries-accumulated scientific mistakes or to remove modern physics from science and to introduce his breakthrough scientific discoveries. Science is as clear as black and white due to that the correct science always stands tall supported by all experimental findings. Now it is clear that the mechanical physics–based science is the only correct science while modern physics is not science. It is time for the general public join the scientific revolution to save our science and science education.

CPSIA information can be obtained
at www.ICGtesting.com
Printed in the USA
BVHW081057160922
647219BV00008B/407